Contents

S0-BBX-024

Note to Caregivers:

Throughout this book, many questions are posed to the reader. Some are open-ended and ask what the reader thinks. Discuss these questions with your child and guide him or her in thinking through the possible answers and outcomes. There are also questions posed which have a specific answer. Encourage your child to read through the text to determine the correct answer. Most importantly, encourage answers grounded in reality while also allowing imaginations to soar. Information to help support you as you share the book with your child is provided in the back in the **Additional Notes** section.

Words that are **bolded** are defined in the glossary in the back of the book.

Weather Report

Cloudy. Sunny. Rainy. Snowy. The **weather** might make you hot or cold, wet or dry. It affects what you wear and where you go. But how does the weather work?

In this book, you will learn about weather. You will also learn about air. After all, there would be no weather on Earth without air.

You will use what you learn to solve a mystery. It is the strange case of a missing hat!

Where Did Your Hat Go?

You put your hat on a park bench. Then you go play. It is windy out. When you come back for your hat, it's gone!

How can you find your hat?

Idea 1: Walk in a small circle. Then make your circle bigger and bigger. Soon, you will have searched the whole playground.

Idea 2: Find out which way the **wind** is blowing. Then follow the wind until you find your hat.

How can you tell by this picture that it is a windy day at the park?

Idea 3: Someone may have seen your hat or picked it up. Ask the other people at the park if they have seen it.

Which idea is best? Why?

Materials

- bubble solution
- bubble wands

Finding the Wind

Sit quietly on the ground outside.

Can you feel any wind on your face?

Blow some bubbles.

Which way do the bubbles go?

What happens to your bubbles after you blow them?

Bubbles are more than just fun.
They can help you solve a mystery!

You can't see wind. But bubbles can show you where the wind is blowing.

Think back to the iScience Puzzle. How can blowing bubbles help you find your hat?

What Can You Do with Air?

You can't breathe underwater. But blowing underwater makes bubbles. They are made of air.

You breathe it in. You breathe it out. Air is all around you. Air is a **gas.**

Bubbles that come from a bottle are made of soap. When you make bubbles from a bottle with a wand, you fill soap with air.

How can you blow bubbles in water? Can you do this without getting wet?

The air you breathe can make things happen. For one thing, it can blow out candles on your birthday cake.

Making Air Move

Air can really move. And you can help it go.

One way to move air is to breathe in and out. You can also blow hard or soft. You can fill a balloon with air. And you can spin a pinwheel with it. You can use a fan to move it.

What else do you think can move air?

Air Movement

You cannot see wind. But flying leaves show you that the wind is blowing.

When air moves, it's called wind.

Your bubbles showed you what direction the wind was blowing.

Think about what wind can do outside on a gusty day. What other clues do you think you can use to "see" the wind?

On long trips, airplanes fly above the clouds. Why is the weather always clear above the clouds?

Did You Know?

Jet stream winds blow from west to east high in the sky. They can blow up to 200 miles (322 kilometers) an hour!

Jet streams help planes go faster when they fly eastward.

What would happen to your bubbles in a jet stream? What would happen to your hat?

When people want a big balloon to go up, they start a fire to heat the air inside. When they want the balloon to come down, they let the air cool.

The Heat Is On

Air can be hot or cold. As it gets warmer, it rises.

Try this. Make sure an adult is with you and says it's okay. Then, blow bubbles near a **radiator** or a heater.

What happens to the bubbles?

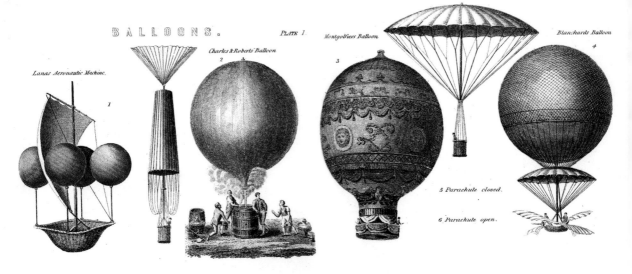

The Montgolfier brothers used hot air to fly in a big balloon. After that, others tried many balloon designs.

Connecting to History

The First Hot-Air Balloon

In 1782, two brothers attached a basket to a big bag. Their names were Joseph and Jacques Montgolfier.

They filled the bag with hot air. It flew!

This was the first hot-air balloon.

The Weather Outside

Clouds are part of the weather. What else is part of the weather?

Weather tells us what is happening in the air outside.

You can find out the **temperature.** This tells you if it's hot or cold.

Clouds can show how much water is in the air.

What else might clouds show?

Meteorologists

Meteorologists study weather. They are scientists. You can see them on your local news talking about the weather. They look into the future. They say if the Sun will shine when you wake up. They **predict** if it might rain this week.

These experts use many **tools.** Thermometers show how hot or cold it is. Rain gauges show how much rain has fallen.

Why would it help to know which way the wind is blowing?

°C °F

50 120
40 100
30
20 80
10 60
0 40
-10 20
-20 0
-30 -20
-40 -40

Knowing the temperature outside can help you decide what clothes to wear.

thermometer

How the Sun Affects Weather

Sunshine makes it fun to play outside. The Sun also plays a big role in all kinds of weather. Its heat warms the air, ground, and water in lakes, rivers, and oceans.

Then, tiny drops of water **evaporate.** That means the water changes from a liquid to a gas. And then it rises into the air.

With help from the Sun, water moves from the ground and the oceans up into the air. There, it makes clouds. What happens next?

Up high, the drops **condense.** That means they turn from a gas back into liquid water. The drops form around little specks of dust. This is how clouds form.

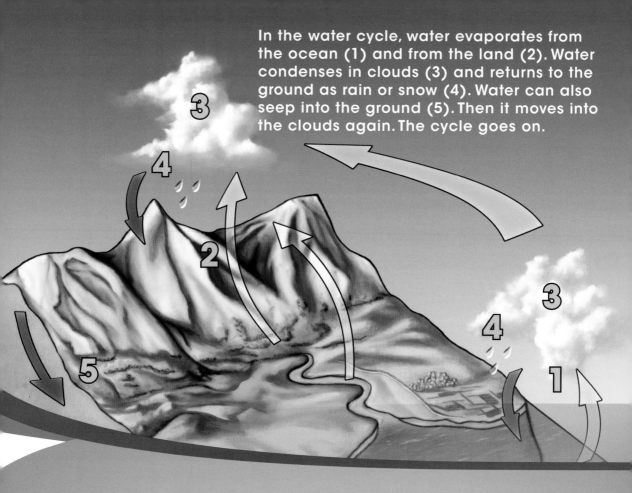

In the water cycle, water evaporates from the ocean (1) and from the land (2). Water condenses in clouds (3) and returns to the ground as rain or snow (4). Water can also seep into the ground (5). Then it moves into the clouds again. The cycle goes on.

Inside clouds, drops of water can get bigger. Then, water falls as rain or snow.

As it falls, water hits the ground and seeps into it. Water can eventually flow into streams, rivers, and even the oceans.

The Sun will shine again. And the **cycle** will keep going.

18

Clouds

Clouds hold water in the air. But not all clouds are the same.

Cirrus clouds are high up and wispy.

cirrus clouds | cumulus clouds | stratus clouds

Different kinds of clouds bring us different kinds of weather.

Cumulus clouds are low and puffy. They look like cotton balls.

Stratus clouds are flat. They appear in layers.

Some kinds of clouds are more likely to bring storms.

What can the wind do to clouds?

Where Did Your Hat Go?

Idea 1 is to walk in circles that get bigger and bigger. This way, you will search the whole park until you find your hat.

Idea 2 is to find out which way the wind is blowing. You could blow bubbles. Then, you could follow them until you find your hat.

Idea 3 is to ask other people if they have seen your hat.

Bubbles helped this boy get his hat back.

If there is a strong wind, idea 2 may work best. It would take you right to your hat. Idea 1 might take longer. But it might help you find your hat on a calmer day. Idea 3 could work. But it won't help you if no one has seen your hat.

Beyond the Puzzle

Now you have learned that air moves. And you used bubbles to tell which way the wind was blowing. You followed the wind to solve a mystery.

How else could you use wind? Read books or look online. Learn about wind power.

Then, design a machine that uses wind to do something. Draw the machine. Show how it uses wind to do work. There's a saying that if you know which way the wind blows, you are informed indeed!

How can looking at the sky tell you about the wind and the weather?

Glossary

cirrus: thin, wispy clouds high in the sky.

condense: change from a gas to a liquid.

cumulus: fluffy clouds that look thick.

cycle: something that happens over and over again in the same order.

evaporate: change from a liquid to a gas without boiling.

gas: something that is not a solid or a liquid.

jet stream: belt of fast-moving air high in the sky.

meteorologists: scientists who study the weather.

predict: to tell what you think will happen at some time in the future.

radiator: a metal box that heats a room when hot water or steam flows through it.

stratus: short clouds that form in layers.

temperature: a measure of the amount of heat.

tools: things to help with a job.

weather: a measure of the air, including how hot it is, how windy it is, and whether it is raining.

wind: moving air.

Further Reading

The Best Book of Weather, by Simon Adams. Kingfisher, 2008.

Air Is All Around You, by Franklyn M. Branley. Collins, 2006.

Wind and Air Pressure, by Alan Rogers and Angella Strelluk. Heinemann Educational Books, 2007.

Weather Wiz Kids, Wind. http://www.weatherwizkids.com/weather-wind.htm

University of Illinois Extension, Tree House Weather Kids. http://urbanext.illinois.edu/treehouse/

Additional Notes

The page references below provide answers to questions asked throughout the book. Questions whose answers will vary are not addressed.

Page 6: Caption question: The girl's hair is moving, so there must be wind.

Page 8: Blowing bubbles and watching where they move might show you in which direction something else, such as a hat, might be moving.

Page 9: You have to put your mouth in water and blow. You could do this without getting wet if you used a straw to blow the air into the water.

Page 12: Bubbles and a hat would probably move very fast in the jet stream.

Caption question: The weather is always clear above clouds because rain and other forms of precipitation come from clouds and move down to Earth, not up into the sky.

Page 13: Bubbles near a radiator or a heater float up.

Page 16: Wind direction can tell you whether storms in another place are heading your way.

Page 17: Caption question: The water comes back down.

Page 19: Wind can move clouds, or break them up, or push them together.

Index